HELLENIC MINISTRY OF CULTURE
ARCHAEOLOGICAL RECEIPTS FUND

VERGINA

Wandering through the archaeological site

STELLA DROUGOU
CHRYSOULA SAATSOGLOU-PALIADELI

ATHENS 2008

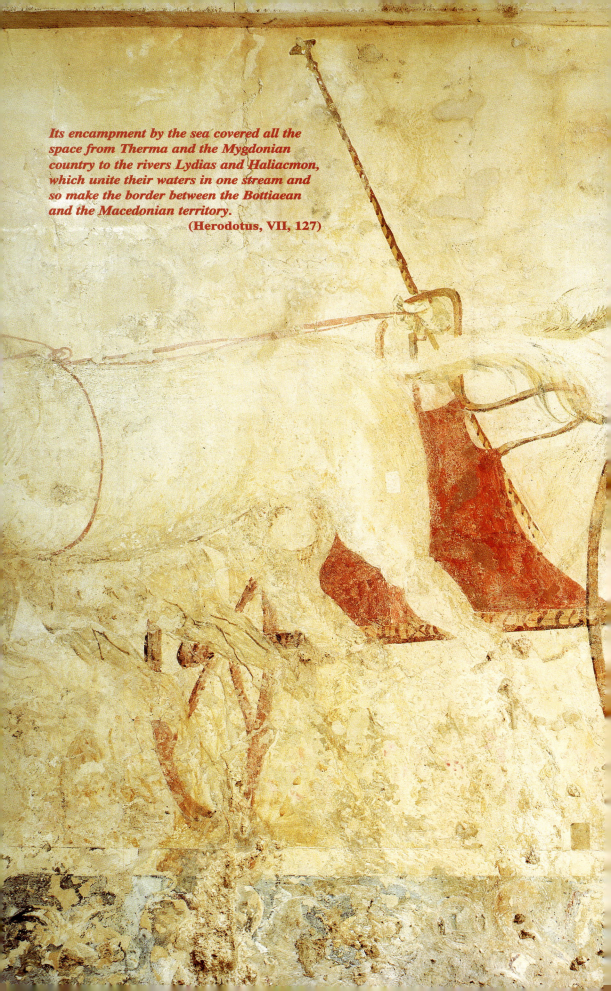

Its encampment by the sea covered all the space from Therma and the Mygdonian country to the rivers Lydias and Haliacmon, which unite their waters in one stream and so make the border between the Bottiaean and the Macedonian territory.
                    (Herodotus, VII, 127)

VERGINA

PREFACE

This book is but a small facet of a large and important project that has been underway at Vergina for many years. It presents a sketch of the vast archaeological site, the monuments in which, some famous, some less well-known, have been a rich new source of knowledge on Greek antiquity and the history of Macedonia. The identification of the ancient city at Vergina as the ancient capital of the Macedonians, Aegae, and the discovery of the royal tombs by Professor Manolis Andronicos, excited lively interest among scholars and laymen alike. Members of the excavation team are now studying the monuments and the issues, and full publications of the former will soon be available to scholarship. However, in our endeavours to continue Manolis Andronicos's tradition of bringing Archaeology to the people, we are also conscious of our duty to heighten public awareness. So, as a follow up to the guide to the Great Tumulus (Megali Toumba), we have prepared this guide to the archaeological site at Vergina as a whole, taking a tour through this fascinating place.

This venture would have been impossible without the continuous support of the Aristotle University, Thessaloniki and particularly of its rectorial authorities. We are grateful to the Ministry of Culture, specifically the Publications Department of the Archaeological Receipts Fund, for enabling us to achieve our true objective, that is to spread information on the archaeological project

to as wide an audience as possible. Of the collaborators on the excavation, Ioannis Graikos, Notis Toufexis, Chrysanthi Kallini spared no effort for the preparation of this book. Most of the photographs, drawings and plans are from the Vergina Excavation Archive, which is kept in the Aristotle University and includes works by such dear colleagues and unforgettable friends as S. Tsavdaroglou, J. Travlos, G. Miltsakakis, N. Haddad, G. Athanasiadis, K. Toutountzidis and G. Gatzios. The most recent photographs of the site were taken by G. Katsangelos. The photographs of the coins were kindly provided by the Numismatic Museum, Athens. The publication owes much to Evangelia Kypraiou and her experienced colleagues at the Archaeological Receipts Fund. We express our heartfelt thanks to all the institutions and individuals mentioned above.

The publication of the guide to Vergina is a token of our esteem for this site and the work that has been achieved there by our teachers and students. The enhancement of the monuments and the communication of what is known about them define the quality of the experience enjoyed by visitors to the archaeological site. This is the goal of our own activity there, with the help of all fellow contributors.

Thessaloniki 1997

Stella Drougou
Chrysoula Saatsoglou-Paliadeli

The geographical location

A few kilometres east of the point where the river Aliakmon (anc. Haliacmon) triumphantly enters the fertile Macedonian plain, separating the Pierian range from Mount Vermion (anc. Bermius), lie the ruins of the old capital of the ancient Macedonians, Aegae. At the other edge of the plain is Mount Paikos, visible when the weather is fine, on the lower reaches of which spreads Pella. The Byzantine village of Palatitsia, situated to the east of the archaeological site, recalls the region's former glory in its name (lit. Little Palace), while Veroia (anc. Beroea) to the northwest, in the foothills of Mount Vermion, preserves beneath its modern houses evidence of its illustrious Byzantine, Roman and Hellenistic past, when the old Macedonian capital had long lost its significance.

The archaeological site

The archaeological site of Vergina comprises two independent units: the ancient city and its cemetery. The ruins of the ancient city spread on the northern slopes of Pieria, while the cemetery extends over the alluvial plain that stretches from the foot of the mountain as far as the Aliakmon.

Well protected by the mountain to the south, as well as by the now dry beds of torrents to west and east, the ancient city was only open on the side overlooking the plain. The naturally fortified site was reinforced by mighty walls that surrounded the settlement completely.

Built on sloping ground, the city was arranged in terraces; on the smallest yet most imposing of which, with unimpeded view over the Macedonian plain, stood the palace. On the heights behind this impressive building are remains of the acropolis, while on the terrace immediately below it, the theatre was discovered in 1982. Even lower down is part of the ancient agora, very close to the west wall of the enceinte.

The extensive cemetery in the lowland area north and east of Vergina

Fig. 1. The river Aliakmon (anc. Haliacmon) as it descends to the Macedonian plain, between Mount Vermion and the Pierian range.

was extra muros of Aegae, as was the case in all ancient Greek cities. The small tumuli (man-made mounds of earth covering one or more burials) can still be discerned right and left of the road linking Veroia and Palatitsia, when the light falls on them obliquely, even though their original form has been much altered by mechanized cultivation. In and amongst the numerous low tumuli of this cemetery, some more prominent mounds covered large, subterranean, sepulchral monuments with vaulted roof and facade, the so-called Macedonian tombs. The systematic excavation at Vergina has brought to light eleven such tombs: one cluster of three is closest to Palatitsia (tombs in the Bella tumulus), another group constitutes the royal tombs of the Great Tumulus, while three more are nearer to the ancient city (the tomb beside the Vergina Cultural Centre, the "Rhomaios tomb" and the adjacent "Eurydike tomb"). The continuous use of the cemetery from 1000 BC till the first century AD not only records a long period of habitation in the region, from prehistoric into Roman times, but also reflects, through the grave goods accompanying the dead, the floruit and decline of the societies living in the Vergina area over more than a thousand years.

The modern village

Until 1922 the site of the present village of Vergina, with its 1.500 or so inhabitants, was occupied by two hamlets, Koutles and Barbes. With the mass influx of refugees that came in the wake of Greece's defeat in the Asia Minor campaign, Greeks from the Caucasus and Pontos settled here. They and the locals made their home in the new village, which was named Vergina after a queen in a fairytale.

In those days the architectural remains of the ancient public and private buildings, in the fields at the foot of the mountain, were better preserved. Their architectural members proved to be handy building material for the refugees, who fortunately decided to build their houses lower down, leaving the greater part of the extensive archaeological site free for research.

Fig. 2. View of Vergina from the palace. The modern village was created after 1922 by settling the inhabitants of the two earlier villages, Koutles and Barbes, and refugees from Pontos and the Caucasus.

During the reign of Napoleon III in France, when Macedonia was still part of the Ottoman Empire, the French archaeologist Leon Heuzey, first located the archaeological site to the west of Palatitsia, in 1856. Impressed by the ruins he beheld, he decided to investigate it. In 1861, after a brief excavation lasting less than forty days, in which he uncovered a Macedonian tomb and the east side of the palace, Heuzey left, taking with him a few stone finds that are nowadays in the Louvre Museum, Paris. In 1876 Heuzey published the results of his research in a volume that has proved invaluable for subsequent archaeologists working at the site.

Heuzey's excavation was resumed some sixty years later by Konstantinos Rhomaios, Professor of Archaeology in the Faculty of Letters at the Aristotle University of Thessaloniki. He selected the area of Vergina as suitable for running a teaching excavation for his students. In 1937 he came to the recently founded village and, together with his students, continued investigating the palace. He also revealed the impressive Macedonian tomb that nowadays bears his name, the "Rhomaios tomb". Among the eager young novices accompanying him then were Ph. Petsas and Manolis Andronicos, whose first practical experience in digging was gained at the site to which he devoted most of his archaeological activity. Such was the inauguration, a few years before the Second World War, the University of Thessaloniki's long association with research at Vergina, which has continued uninterrupted to this day.

The systematic excavation of the extensive tumuli cemetery by Manolis Andronicos (1952-1963) and the large-scale rescue excavation by Ph. Petsas, in the course of laying the Vergina-Palatitsia road, culminated in the excavation of the Great Tumulus and the discovery of the royal tombs (1976-1980). The following year (1981) investigations concentrated on the area of the ancient settlement, that is the acropolis and the terraces below the palace. Although less spectacular than the precious finds from the royal tombs, the achitectural remains of the ancient city nonetheless shed ample light on the history and civilization of the ancient Macedonians, amplifying our knowledge of the identity of a people regarded with hostility by its contemporaries and suspicion by later historians.

Fig. 3. Map of the archaeological site and the surrounding area. The ancient city spread on the north foothills of the Pierian range, in the heart of Herodotus' Macedonian territory. (Plan by G. Gatsios - A. Saayah).

LEGEND
1. The Palace
2. The Theatre
3. The Sanctuary of Eykleia
4. Public Buildings
5. The Rhomaios Tomb
6. The Hellenistic House
7. The Sanctuary of the Mother of the Gods
8. The Great Tumulus
9. The Tumuli cemetery
10. The Tombs of the Bellas farm
11. The Acropolis

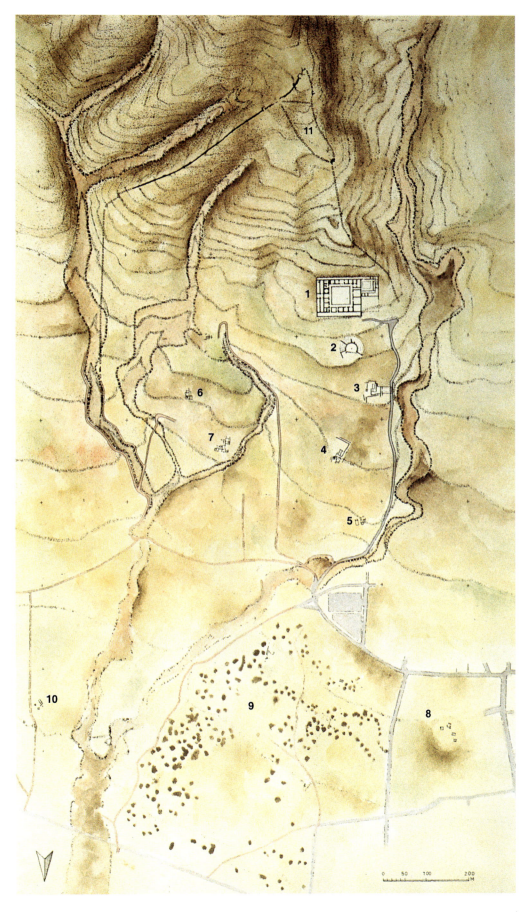

THE EXCAVATIONS

Manolis Andronicos's excavations in the extensive tumuli cemetery, between 1952 and 1963, and his exemplary publication of the finds in 1969, showed that the region was inhabited continuously from 1000 BC until the first century AD. Enjoying a sustained heyday for the greater part of this long period, as attested by the impressive finds of the early settlement phase (1000-700 BC) as well as of Archaic, Classical and Early Hellenistic times (6th-3rd century BC) —from both the cemetery and the fortified settlement—, the city fell into decline after Macedonia's conquest by the Romans, primarily after Andriskos' failed revolt in 148 BC. Signs of this degenerative course, which was completed in the mid-first century AD, are apparent in the later building phases of the ancient settlement and in the poor graves of this period, at the end of which the city was abandoned. Its inhabitants sought other places in which to settle anew.

THE IDENTITY OF THE ANCIENT CITY

The first excavator of the archaeological site, Leon Heuzey, had proposed its identification with the insignificant Macedonian city of Ballas, despite the many reservations that the obvious high quality of the architectural remains, the presence of the imposing palace and the extensive cemetery created. This identification was accepted, for no particular reason but with the same reservations, by later researchers at the site.

In 1968 the eminent British historian Nicholas Hammond, an expert on Macedonian topography, made a revolutionary proposal: that the archaeological site at Vergina was in fact Aegae, the first capital and royal necropolis of the ancient Macedonians. Basing his hypothesis mainly on topographical and historical arguments, Hammond considered that the impressive palace and the then few but significant underground sepulchral buildings (the "Heuzey tomb" and the "Rhomaios tomb", named after the archaeologists who discovered them), in conjunction with the extensive tumuli cemetery, constituted not only proof of the importance of the archaeological site at Vergina but also credible clues to its identification as the cradle of the kingdom of Macedon. Aegae, in the heart of the "land of Macedonia", to quote Herodotus, was, according to the meagre literary tradition, the starting point for the rapid territorial expansion of the kingdom of Macedon. The site was chosen by the founder of the Macedonian dynasty, Perdiccas I, as his capital, following a prophecy of

Fig. 4. Map of the archaeological site as published by L. Heuzey - H. Daumet in their invaluable book, Mission archéologique de Macédoine, *1876.*

the oracle at Delphi. From here his successors began their predatory activity, as described by Thucydides, subjugating to their realm Pieria to the southeast, Bottiaea to the northwest and Mygdonia to the north, already before the end of the sixth century BC.

When King Archelaos transferred the capital to Pella, in the late fifth century BC, the city of Aegae lost none of its significance. It retained its role as the royal burial ground and received the members of the

Fig. 5. Silver tetradrachm of Philip. Pella mint (356-348 BC). Numismatic Museum, Athens.

This Alexander was seventh in descent from Perdiccas, who got for himself the despotism of Macedonia in the way that I will show. Three brothers of the lineage of Temenus came as banished men from Argos to Illyria, Gauanes and Aeropus and Perdiccas; and from Illyria they crossed over into the highlands of Macedonia till they came to the town of Lebaea. There they served for wages as thralls in the king's household, one tending horses and another oxen, and Perdiccas, who was the youngest, the lesser flocks. Now the king's wife cooked their food for them; for in old times the ruling houses among men, and not the commonalty alone, were lacking in wealth; and whenever she baked bread, the loaf of the thrall Perdiccas grew double in bigness. Seeing that this ever happened, she told her husband; and it seemed to him when he heard it that this was a portent, signifying some great matter. So he sent for his thralls and bade them depart out of his territory. They said it was but just that they should have their wages ere they departed; whereupon the king, when they spoke of wages, was moved to foolishness, and said, 'That is the wage you merit, and it is that I give you,' pointing to the sunlight that shone down the smoke vent into the house. Gauanes and Aeropus, who were the elder, stood astonished when they heard that; but the boy said, 'We accept what you give, O king,' and with that he took a knife that he had upon him and drew a line with it on the floor of the house round the sunlight; which done, he thrice gathered up the sunlight into the fold of his garment, and went his way with his companions.

So they departed; but one of them that sat by declared to the king what this was that the boy had done, and how it was of set purpose that the youngest of them had accepted the gift offered; which when the king heard, he was angered, and sent riders after them to slay them. But there is in that land a river, whereto the descendants from Argos of these men offer sacrifice, as their deliverer; this river, when the sons of Temenus had crossed it, rose in such flood that the riders could not cross. So the brothers came to another part of Macedonia and settled near the place called the garden of Midas son of Gordias, wherein roses grow of themselves, each bearing sixty blossoms and of surpassing fragrance; in which garden, by the Macedonian story, Silenus was taken captive; above it rises the mountain called Bermius, which none can ascend for the wintry cold. Thence they issued forth when they had won that country, and presently subdued also the rest of Macedonia.

Herodotus VIII, 137-138

Fig. 6. Silver stater of Alexander I. Amphipolis mint (c. 330-320 BC). Numismatic Museum, Athens.

Fig. 7. Gold ornament in the form of a gorgoneion, from the antechamber of Philip II's tomb.

royal family on those special occasions, such as weddings and funerals, when custom decreed their presence in the old capital. This explains the testimony in the ancient sources that Philip II was assassinated at Aegae, during the magnificent ceremony in the city's theatre to celebrate his daughter's marriage to King Alexander of Epirus in 336 BC, as described by Diodorus Siculus (XVI, 92-94). The heinous deed was witnessed by the heir to the Macedonian throne, the youth Alexander, who punished the assassin and all those involved in the plot. Declared King of Macedon right away, after the approval of the Macedonian army, he fulfilled his filial obligations with a splendid funeral for his father at Aegae. He was now ready to embark on his great expedition to the East.

The archaeological evidence

Nicholas Hammond's proposed identification of the archaeological site at Vergina with the royal cemetery of the ancient Macedonians gained credence in 1976, thanks to the archaeological evidence from the systematic excavation of the Great Tumulus by Manolis Andronicos and his collaborators. It was confirmed one year later with the discovery of the

The archaeological investigation of the area has so far provided copious material both for our knowledge of the Macedonians and for our understanding of their historical physiognomy. Its continuation, and the systematic study of the finds, can only assist us to draw a much clearer and more accurate picture of this domain of Hellenism which from the fourth century onwards play a leading role in the historical process and succeeded in spreading the Greek language and Greek culture over virtually the whole world.

M. Andronicos, *Vergina. The Royal Tombs and the other Antiquities*, 1984

unlooted royal tombs. Their monumental architecture, but above all their remarkable painted decoration with subjects such as the lion hunt, which are associated directly with the royal family, as well as the superb quality of the objects accompanying the dead, validated Hammond's theory.

The extension of the excavation to the area of the ancient city brought to light the theatre of Aegae, in 1982, exactly below the palace and directly related to it in the urban tissue, so bringing to life Diodorus Siculus' description of the assassination of Philip.

The location of the sanctuary of Eukleia a little below the theatre, its identification with part of the agora of Aegae and the discovery of inscriptions preserving the name of Eurydike, daughter of Sirras, wife of Amyntas III and mother of Philip II (and therefore grandmother of Alexander), left no doubt as to the character of the archaeological site at Vergina. The obvious royal presence in the palace (the provenance of two inscriptions dedicated to Herakles Patroos, genarch of the Argeades dynasty), the theatre, the sanctuary of Eukleia and the cemetery, demonstrate that the wider archaeological site of Vergina corresponds to the old capital and cultural cradle of the ancient Macedonians. The impressive palace, the theatre and the agora, the sanctuaries of Eukleia and the Mother of the Gods, a large Hellenistic house and other private and public buildings, the eleven Macedonian tombs to date, the extensive prehistoric tumuli cemetery, the Archaic tombs with rich grave goods and the funerary and dedicatory inscriptions of Classical

Fig. 8. Inscribed and painted funerary stele from the fill of the Great Tumulus at Vergina.

and Hellenistic times are precious evidence for reassessing the history and culture of the ancient Macedonians. The archaeological remains of their old capital reflect aspects of the private and public life of its anonymous and eponymous inhabitants that supplement in an objective manner the few testimonies in the ancient sources.

Chr. S.-P.

THE ANCIENT CITY

The ancient city was girt by a wall which is nowadays neither visible nor accessible to visitors. The section that surrounded the area of the acropolis, at the highest point to the south of the city, is one of the best preserved, along with that to the east, where the gateways with their circular towers survive.

The ancient city is still dominated by the palace and the theatre, a short way to the north, both monuments which can be seen by the visitor. Very close to these public buildings, on the same side of the city, is its political hub, the agora, where the sanctuary of Eukleia was found with important marble statues, ex-votos of Philip's II mother, Eurydike.

On the east side of the city, an important sanctuary dedicated to the Mother of the Gods-Cybele, with characteristic finds, has come to light in recent years. This monument cannot be visited at present.

At various points in the ancient city private houses and other buildings have been excavated, which are still only accessible to the archaeologists (buildings in the Efraimidis field and house in the Tsirelas field).

THE FORTIFICATION WALL AND THE ACROPOLIS

L. Heuzey, in his investigations in the nineteenth century, had ascertained remnants of the ancient city wall; he also mentions the existence of a gate. These remains are recorded in plans drawn by H. Daumet (L. Heuzey - H. Daumet, *Mission Archéologique de Macédoine*, 1876). Recent excavations by P. Faklaris have confirmed the French researchers' finds and provided much new evidence: the west leg of the wall rose immediately west of the palace and above the river, reaching the summit of the highest hill of the city at the point where the triangular area of the acropolis is formed. Gradually traces of the north wall are being brought to light. It then continued, following the gradient of the ground and the course of a large torrent in its eastern section, and descended to the plain, so circumvallating the large ancient city. The north leg of the wall is still unknown.

The gateways with their circular towers were uncovered recently on the east side. Careful masonry as well as signs of diverse interventions belie the importance of the Hellenistic fortification of the city. During recent years other parts of the east wall were brought to light. In the area of the acropolis, remains of workshops and their activities have been revealed, indicating this sector's important role.

Fig. 9. The large building of the palace with the central peristyle court occupies a commanding position on the west side of the ancient city. It stands on the wide terrace, protected from the west leg of the fortification wall and the deep river on the same side. Visible at a lower level is the theatre of Aegae (4th century BC).

THE PALACE

The palace at Vergina dominates, by virtue of its position and size, the ancient city of Aegae. Built on a large terrace on its west side, protected from the deep ravine to the west and the walled acropolis to the south, it constitutes the principal building in a large urban complex that includes the theatre, the agora with the sanctuary of Eukleia and perhaps other buidings that evidently existed hereabouts. It is thus obvious that the west side of the city, with all these edifices, was the centre of civic and political life.

In the history of the excavations at Vergina the palace was the first monument recognized in the area, which was known already from the mid-nineteenth century thanks to L. Heuzey. Later, it was the focus of the excavations conducted by the University of Thessaloniki until 1975. In 1861 the French archaeologist and traveller investigated mainly the east part of the building. The palace was uncovered completely in the twentieth century, with the excavations of professors K. Rhomaios, G. Bakalakis and M. Andronicos. The large building occupies the extensive west terrace of Aghia Triada, while at its feet, north and east, spreads the ancient city which occupies an area of about one hectare (104.50 × 88.50 m). The architecture of the palace essentially reproduces that of the ancient Greek house with the central peristyle court. Its plan is remarkable for the simplicity of its conception, since everything is organized around a spacious central court surrounded by four large Doric porticoes. Outstanding is the court's well-built poros gutter, in which rain water from the portico roofs was collected. On the south side are huge rooms with wonderful mosaic floors, the formal halls of the palace where the famous Macedonian drinking parties (*symposia*) were held. Particularly interesting is the suite of four large rooms arranged in pairs, either side of an open space which communicates with the south side of the large court. The floors of the rooms were covered with mosaics with impressive palmette patterns.

That in room E merits special attention, with its white, black, red and grey tesserae embedded in plaster, in a lavish composition combining rhythm, decorativeness and an attempted naturalism, while most probably enclosing a distinct and religious symbolism: at the centre is a double rosette from which sprout pairs of tendrils embellished with flowers and leaves, all set against a black ground. The rich vegetal composition is encircled by bands of meander pattern and spiral meander. At the corners, in the small triangular spaces, female floral figures, richly bedecked with earrings and polos on the head, hold branches with flowers and leaves, symbolizing the omnipotence of nature.

The palace not only housed the members of the royal family, but also friends, companions (*hetairoi*), visitors and warriors. The enor-

Fig. 10. View of the palace at Vergina. Built on a wide terrace on the west side of the ancient city, it surveyed the Aliakmon plain. The formal rooms were located on its south side.

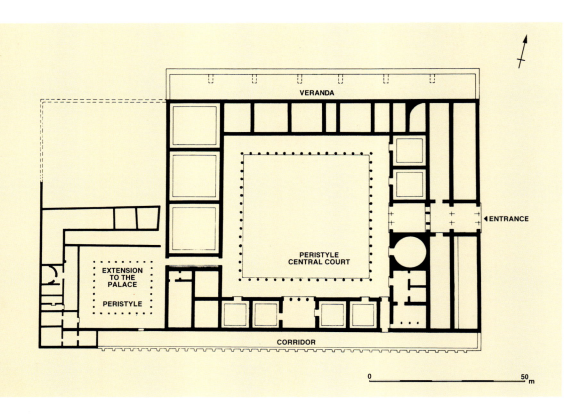

VERANDA

PERISTYLE
CENTRAL COURT

ENTRANCE

EXTENSION
TO THE
PALACE

PERISTYLE

CORRIDOR

0   50 m

Fig. 11. The concept of the architectural plan of the palace is based on that of the 4th-century BC Greek house. The large, central, peristyle court is surrounded by a Doric colonnade onto which the main rooms open. On the east side there was an upper storey, as the Ionic architectural members indicate, and a circular chamber, the tholos, dedicated to Herakles Patroos. The entrance to the palace on that side was enhanced by its monumental aspect. (Plan by I. Travlos - D. Pandermalis).

mous rooms on the west side of the building evidently accommodated a lot of people, apparently in luxury, as their *opus sectile* floors indicate. Impressive was the magnificence and the monumentality of the entrance on the east side, with its double chambers and large pristine white thresholds. There was a second storey on this side, in which the king seems to have had his private apartments.

The visitor to the palace first passed through its gateway and then, before coming into the large court, entered the most sacred place in the whole building, the tholos. In this circular room with a mosaic pave-

ment, Herakles Patroos was worshipped, as the inscription incised on a marble relief plaque found inside the tholos attests. The north side of the palace was particularly important. There, on a mighty podium, ran a long veranda, a special feature, unprecedented in the architecture of the Greek house, from where the king and his entourage could survey the city and the Aliakmon plain, northwards as far as the eye could see.

The impressive marble thresholds, the mosaic pavements, the Doric and Ionic colonnades reveal the opulence of the splendid building as well as the lifestyle enjoyed by

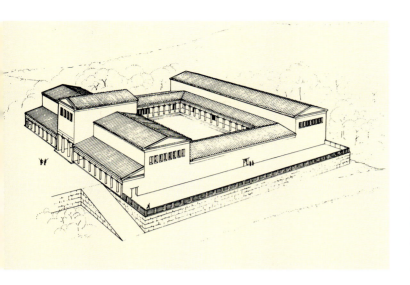

its residents. The large relief plaques now in the Louvre, together with other finds from L. Heuzey's excavation, come from the palace. In addition, the stone conduits and the elaborate roof tiles, as well as the various functional architectural members, all of exceptional workmanship, completed the building's grand appearance. Furthermore, the individual architectural features (corridors, staircases, conduits) reveal its functionality. Characteristic examples of the skill of the masons and craftsmen employed on its construction are the three large rooms in the west wing (16.74 x 17.66 m) with their lovely white *opus sectile* and zones of black tessellated pavement. These halls must have been used for special functions, such as the wonderful banquets atended by many guests. The excavation showed that the walls were built of brick on a poros socle and coated with very fine quality stucco with alternating red, white, yellow and black surfaces. It is almost certain that this sumptuous building was also adorned with murals and painted compositions, as attested in Macedonia from the texts and revealed by monuments with similar use and function, such as the palaces and houses at Pella and the richly decorated Macedonian tombs.

Recent archaeological research has shown that the palace must have been built together with the theatre in the second half of the fourth century BC. Some changes were made at a later date and it was finally abandoned in the second century BC, when the reign of the Antigonids was dissolved in Macedonia and the Romans assumed sovereignty.

The house on the west side of the palace was constructed in Hellenistic times, and indeed with building material in secondary use. Of the same architectural type as the palace, with a central peristyle court, it includes mainly ancillary rooms, kitchens and storerooms, and presumably serviced the domestic needs of the inhabitants of the palace. It was deserted and destroyed at the same time as the palace complex.

Figs 13-14. Outstanding in the palace were the large halls for symposia (male drinking parties), with mosaic floors and marble thresholds. Particularly impressive is the mosaic with circular vegetal composition, in room E. In the four spandrels anthokorai (floral maidens) hold blossoms and shoots, expressing the power of nature. The pavement was surrounded by a narrow ledge on which the symposium couches were placed.

Fig. 15. One of the large composite Ionic column capitals of the half-columns on the monumental entrace to the palace (Louvre Museum).

Fig. 16. Three large Ionic double-sided half columns delimit the entrance to the luxurious symposium halls with their marble thresholds and mosaic floors. The open entrance with the double-sided half columns faces onto the large central court.

Fig. 17. As the visitor passed through the large entrance on the east side of the palace, he beheld the tholos, a circular area dedicated to Herakles Patroos.

The ancient theatre at Vergina was discovered in 1982. It lies at the base of the large palace terrace, which fact suggests that the two buildings were conceived as a single complex with related and reciprocal functions and uses. For this reason the presence of other public buildings and installations in their immediate vicinity, can no longer be ruled out, since the entire area apparently had a special public and political character. This would explain the choice of the site for the theatre, even though it does not seem to be the best. The theatre's cavea, facing north, rests against the hillside only in its east part. On the contrary, its west part was probably formed with the help of wooden bleachers, which means that the seats were wooden in some cases. Only the seats of the first row (*edolia*) around the orchestra were of stone. Eight large stone-paved passages 0.74 m wide divide the cavea into nine cunei. The *parodoi*, the entrances to the orchestra, 15 m long, were on the east and the west side of the theatre. The east entrance must have been especially important, since the road leading down from the palace ended there. Commencing at the central entrance of the palace, and perhaps passing through stoas, the road terminated at the east entrance to the theatre, where a small court emphasizes its special character. The section of the corridor in the rock, revetted with rectangular, dressed poros slabs to enhance its monumentality, together with another wall form the sep-

Fig. 18. The ancient theatre. This is perhaps one of the earliest stone theatres in the ancient Greek world, since it was probably designed and built in the second half of the 4th century BC. The stone parts are constructed of poros. Earlier building remains reveal that its function was perhaps similar in its previous phase.

Fig. 19. The stone conduit on the periphery of the orchestra, cleverly arranged to channel the rain water out of the theatre.

agora, to the north of the theatre. The roads from the palace to the theatre and from the agora to the palace and the theatre, connected the main buildings and complexes in the west part of the city, which thus comprised its political centre. In front of the stone seats in the first row, on the periphery of the orchestra, is the gutter of the theatre, an open channel 0.50 m wide, that carried off the rain water to the west, outside the orchestra and the theatre. The well-carved poros *edolia*, the clear section of the drainage

arate court. The corresponding west parodos, on the other side, is constructed differently. It is entirely built, since there is no rock to support the construction. Perhaps this was the terminus of the road leading from the sanctuary of Eukleia and the

Straightway he (i.e. Philip) *set in motion plans for gorgeous sacrifices to the gods joined with the wedding of his daughter Cleopatra, whose mother was Olympias; ...*
So great numbers of people flocked together from all directions to the festival, and the games and the marriage were celebrated in Aegae in Macedonia. ...
Every seat in the theatre was taken when Philip appeared wearing a white cloak, and by his express orders his bodyguard held away from him and followed only at a distance, since he wanted to show publicly that he was protected by the goodwill of all the Greeks, and had no need of a guard of spearmen. Such was the pinnacle of success that he had attained, but as the praises and congratulations of all rang in his ears, suddenly without warning the plot against the king was revealed as death struck ...
He (Pausanias, the assassin) *posted horses at the gates of the city and came to the entrance of the theatre carrying a Celtic dagger under his cloak. When Philip directed his attending friends to precede him into the theatre, while the guards kept their distance, he saw that the king was left alone, rushed at him, pierced him through his ribs, and stretched him out dead; then ran for the gates and the horses which he had prepared for his flight. Immediately one group of the bodyguards hurried to the body of the king while the rest poured out in pursuit of the assassin; among these last were Leonnatus and Perdiccas and Attalus. Having a good start, Pausanias would have mounted his horse before they could catch him had he not caught his boot in a vine and fallen. As he was scrambling to his feet, Perdiccas and the rest caught up with him and killed him with their javelins.*

Diod. Sic. XVI, 91-94

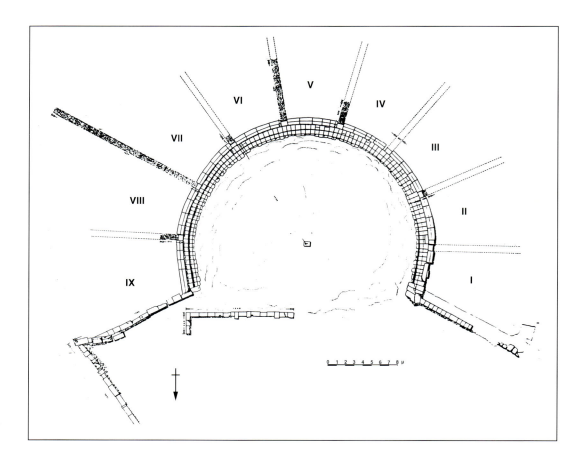

Fig. 20. The ancient theatre lies almost exactly below the north veranda of the palace. It was built on a low hill slope, overlooking the city's agora to the north. This location also determined its additional uses, apart from theatrical performances, as inferred from the testimony of the wedding of Philip II's daughter Kleopatra, on which occasion the monarch was assassinated in 336 BC.

channel, the well-built walls of the *parodoi* reveal a special care that is only observed in the palace. This indicates the unity of the two buildings, a unity and similarity that perhaps point to their founding contemporaneously, in the second half of the fourth century BC, when they perhaps replaced earlier analogous structures. If all these suppositions are correct, then a significant ascertainment can be made about the theatre at Vergina. It is not only one of the largest theatres in the Greek world but also one of the earliest in which stone was used in various parts.

The orchestra, the large open space for the actors between the *parodoi* and the cavea, is shaped like a very open horseshoe, closed to the north by the skene. It is 28.40 m in diameter, while at the centre of the earth floor is preserved the stone base of the altar to Dionysos, the *thymele*, which was perhaps wooden and portable, to be moved as the needs of cult or the occasion demanded. The remains of the skene are scant and its original form is difficult to reconstruct. Only the foundation (surviving length 12.50 m) of the east half of a long narrow building is visible. This most probably formed

Fig. 21. The east section of the conduit with its blind edge. Part of the one and only row of stone seats (edolia) is visible.

openings towards the orchestra, presumably to support the necessary scenery depending on the play performed. The few remnants of walls that have been noted below the foundations of the skene belie the existence of an earlier building that was covered by the skene building. Perhaps these are remains of an earlier and simpler theatral installation, indeed with a different orientation.

The presence of the palace, as well as of the agora with the sanctuary of Eukleia further north, dictated the location of the theatre in this part of the city, to complement these public buildings. So the dramatic event of Philip's II assassination in the theatre at Aegeae on his daughter's wedding day is almost tangible in its setting and the descriptions of the ancient authors come alive.

S.D.

It is not easy to put an exact date on this theatre. However, from such evidence as we have I believe that it can be dated to the fourth century BC. It is worth adding that, aside from its exact date, we may be certain that it was within the confines of this theatre that Philip was murdered in 336 BC since the site of the theatre did not change from one period to another. This alteration would not have been possible, firstly because there is no other area easily available which was suitable for the construction of such a building, and secondly because the site of a theatre was endowed with a certain sanctity linked with the worship of Dionysos.

M. Andronicos, *Vergina. The Royal Tombs and the other Antiquities*, 1984

The site and its significance

About half way between the "Rhomaios tomb" and the palace, to the left of the tarmac road that follows the course of the west branch of the defensive wall, lies the sanctuary of Eukleia, which was excavated in 1982. This part of the archaeological site at Vergina, inside the boundaries of the fortified settlement, was named conventionally from two inscriptions preserving the name of the panhellenic goddess Eukleia. The fact that this goddess was worshipped in the agoras of ancient Greek cities indicates that her sanctuary at Vergina was in the agora of Aegae. Its relation to the theatre and the palace in the urban tissue, the public character of its architectural remains and the royal ex-votos found there confirm that this site was the political centre of the old Macedonian capital.

Fig. 22. Plan of the sanctuary of Eukleia at Vergina. The most important building phase dates to the 4th century BC, to which belong the three marble ex-votos (1-3), a two- room temple (4), the foundation of a large altar to the east (5), a tripartite stoa (6) and the foundations of a building with internal peristyle court (7).

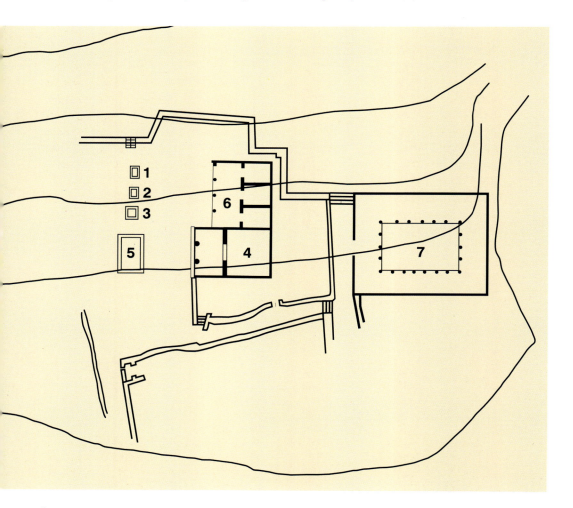

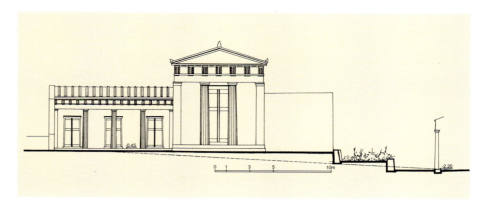

Fig. 23. Reconstruction drawing of the front of the sanctuary of Eukleia in the 4th century BC. (Reconstuction drawing by N. Haddad).

The architectural remains

The excavation in this part of Aegae has so far brought to light the foundations of two temples, an altar, a stoa, a peristyle building, marble pedestals for ex-votos to the deities worshipped here and the remains of a large retaining wall that delimited the southern edge of the large terrace on which the sanctuary developed.

Most of these buildings date from the time when the kingdom of Macedon was at its zenith, during the reigns of Philip and Alexander.

Parts of two road axes, one of which traversed the agora while the other led towards the theatre, help our understanding of the layout of the ancient city, which was built on terraces adapted to the natural declivity of the ground towards the plain.

The finds and their significance

In the two inscriptions from the sanctuary preserving the name of Eukleia, the name of the woman who made a dedication to the goddess is preserved as well. She was Eurydike, daughter of Sirras, who is identified thanks to the ancient texts as the consort of Amyntas III, mother of Philip and grandmother of Alexander. These inscriptions,

Fig. 24. Restoration drawing of a colossal marble snake from the sanctuary of Eukleia. Its original height was 1.80 m. (Drawing by G. Miltsakakis).

Fig. 25. The portrait head of the statue dedicated by queen Eurydike, mother of Philip II, in the sanctuary of Eukleia (see also Fig. 26). The facial features of the mature woman perhaps reflect those of the queen herself.

a royal presence in this part of the ancient city as well. The marble sculptures unearthed in the sanctuary are of great value for our knowledge of ancient Greek art and religion. Among them are part of a colossal snake, of total height 1.80 m, representing some deity in serpentine form, maybe Zeus Meilichios, two marble heads from two statues of the fourth century BC and the complete marble statue of a female figure dedicated by queen Eurydike to the goddess Eukleia, also dated to the fourth century BC. The head of the statue, with the features of a mature woman, perhaps represents the queen herself, in which case this is the unique portrait of Eurydike to have survived from antiquity.

The presence of leading intellectuals and artists in Macedonia is known from the ancient sources, which mention the dramatist Euripides, the painter Zeuxes and the musician Timotheos. It is obvious too from the marvellous murals adorning the royal tombs in the Great Tumulus. Now it is confirmed by other

together with a third one from the base of a statue of the queen, found in nearby Palatitsia, are the only ones from antiquity mentioning Eurydike.

Precious as historical testimonia, they also reveal the importance of their find spot, because they denote

Fig. 26. The marble statue dedicated by queen Eurydike to the goddess Eukleia. It was found buried next to the foundations of the large temple. The female head in Fig. 25 belongs to this impressive sculpture, 1.90 m high, from the middle years of the 4th century BC.

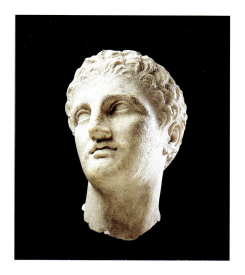

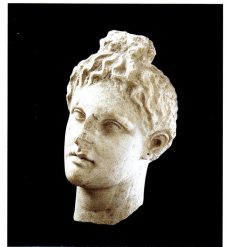

Fig. 27. Marble male head of the 4th century BC, from the Sanctuary of Eukleia. It belonged to a statue of a young god, hero or heroized mortal.

Fig. 28. Marble female head from the sanctuary of Eukleia. It perhaps belonged to a statue of the young goddess (4th century BC).

works, namely the sculpture from Aegae. So the finds from the sanctuary of Eukleia reinforce the view that despite the transfer of the administrative centre of the Macedonian kingdom to Pella, the old capital maintained its importance and splendour. This is what we are trying to reconstruct on the basis of the remains that the hospitable soil of Vergina has preserved for its excavators, for archaeological science and for visitors to this intriguing place.

Ch. S.-P.

Fig. 29. Inscribed pedestal of an ex-voto of unknown form, dedicated by queen Eurydike, mother of Philip II, in the sanctuary of Eukleia.

Fig. 30. The pedestal of the statue dedicated by queen Eurydike, in the sanctuary of Eukleia. Its inscription, identical to that in Fig. 29, declares the royal presence in the site, which is identified as part of the agora of Aegae.

I ate from a drum, I drank from a cymbal, I brought offerings, I emerge from under the bridal bed.

These are the characteristic words spoken during the initiation rites for the devotees of the archaic deity, the Mother of the Gods, who was worshipped in Greece and particularly in Macedonia with fervent piety. Her cult was linked with arcane mysteries, involving special observances for the devotees. Its close relationship with Dionysiac cult is readily apparent, since orgiastic dances by women, a state of trance and loud music were essential components of the veneration of both deities. By the early sixth century BC the Mother of the Gods was linked iconographically as well as in worship with her Asian counterpart, Cybele, goddess of the seasons, of cities and of fortification walls.

The remains of the *metroa* (sanctuaries) in which this has been noted in excavation in Greece, in parallel with the information in the ancient authors, reveal the importance of the cult of this great female deity. The poet Pindar founded the Metroon at Thebes, near his home, where women worshipped the goddess with orgiastic dances and music. The image of the goddess is preserved mainly in votive reliefs and terracotta figurines with typical traits:

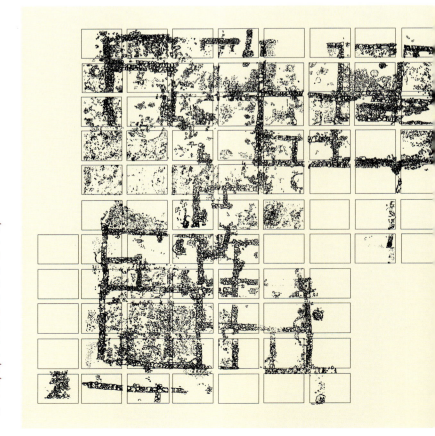

Fig. 31. The Metroon of ancient Aegae, at Vergina, is one of the most important sanctuaries of the Mother of the Gods Cybele. It is a large and composite complex of places for cult as well as other anciellary rooms. Built in the early 3rd century BC on the ruins of the earlier Metroon of Classical times, the Hellenistic sanctuary was destroyed c. 150 BC.

she is seated on a throne, sometimes embracing and sometimes stepping on the divine animal, the lion cub. In her right hand she holds a libation bowl and in her left the large drum of the orgiastic dances. On her head, above the elaborate coiffure, she wears either a high mural crown or a narrow diadem. Other deities were venerated with her, co-regnants or consorts, such as Sarapis, Aphrodite and Persephone, who formed a coterie (*synteleia*) of religious and cult beings accompanying the Great Goddess. The Metroon at Vergina came to light in 1990 and from the outset — and in the following years — its remains yielded a wealth of finds and information on the worship of the Mother of the Gods at Aegae and in Macedonia in general. By good fortune a kantharos, a clay wine cup, incised with the inscription: *ΜΗΤΡΙ ΘΕΩΝ ΚΑΙ ΣΥΝΤΕΛΕΙΑΙ* (Mother of the Gods and companions), was discovered in a small room in the sanctuary, in the very first season. So the deity to whom this large an d peculiar sanctuary was dedicated was identified with certainty.

In recent years a large quadrilateral edifice (32 × 32 m approx.) has

Fig. 32. Terracotta figurine of the Mother of the Gods with the characteristics of the Phrygian goddess Cybele. Seated on a throne, she holds a drum in one hand and a bowl in the other, while her legs rest on her sacred animal, the lion (2nd century BC).

Figs 33-34. The dedicatory inscription in honour of the Mother of the Gods, incised on a black-glaze kantharos (c. 300 BC), enabled the recognition and identification of the main deity in the sanctuary.

been uncovered gradually. It includes large roofed areas where devotees assembled for their initiation and worship, auxiliary areas and secluded courts with several altars-eschars for burnt sacrifices in honour of the goddess. At the centre of the large, plain halls with earth floors were large hearths which will have provided heat and light at the time of initiation and worship. Noteworthy is the large double room on the southwest side and the very important room in which large terracotta figures of the goddess and companion divinities were stored. The presence of deities such as Zeus-Sarapis, Aphrodite and Persephone, and primarily the ritual paraphernalia from the East, such as the bone-

Fig. 35. Head of a terracotta statue of Cybele. The goddess, shown with the features of Aphrodite, wears a narrow diadem in her hair and earrings (late 3rd-early 2nd century BC).

Fig. 36. Terracotta nude female figure from the Metroon. It is interesting that wings sprout from the shoulders. Perhaps an unknown daemonic figure associated with Aphrodite.

handled keys which were unexpectedly found in the same room as the terracotta figures, underline the meaning of the mystery cult and belie the importance of the occult and the initiation for the devotee entering the sanctuary.

Outstanding among the many finds

Fig. 37. Terracotta figurine of a bull and model of a festooned altar, both characteristic finds from the sanctuary of the Mother of the Gods.

Fig. 38. One of the terracotta figures of the goddess as found among the roof tiles of the Metroon.

from the sanctuary of the Mother of the Gods are the terracotta figures of the goddess in the familiar type of Cybele, the censers in various shapes, other cult vessels and the terracotta animal figurines, all interesting examples of local coroplastic art. Together with these a host of vases and coins, loom weights and many architectural members from earlier buildings make up a very rich assemblage of objects associated with both life and worship in the sanctuary. The construction of the large building was so simple that it could be characterized as vernacular. It had brick walls and only the socle was of stone. The floors were earthen and the roofs must have been low, in harmony with the gentle slope of the landscape. Dated to Hellenistic times, it was destroyed by conflagration around 150 BC. The excavation has shown, however, that it was built on the ruins of an earlier building of the fourth century BC, proof that the cult of the goddess at Vergina already existed in Classical times and continued with great intensity in Hellenistic, as has been ascertained both in Macedonia and the rest of the Hellenistic world.

Fig. 39. Clay loom-weights were apparently common votive offerings in the sanctuary of the Mother of the gods. A large ensemble was found next to the entrance of the sacred room of the "keys".

The extensive cemeteries of the ancient city of Aegae at Vergina span a long period, almost ten centuries. They spread on the plain north of the city, in the area between the present villages of Palatitsia and Vergina. A large section of the cemetery to the east dates to the early centuries (10th-7th century BC) when the kingdom of Macedon was taking shape. The centuries of the kingdom's rise and heyday (6th-4th century BC) are represented by rich and very important sepulchral monuments, the majority on the outskirts of the present village of Vergina. Large Archaic tombs as well as Early Classical Macedonian tombs from the fourth century BC create one of the major groups of monuments in Macedonia with their architecture, painting and grave goods. They bear witness to the kingdom's floruit under the leadership of such great kings as Perdiccas I, Alexander I, Archelaos, Philip II, Alexander III and Cassander. Many simple tombs as well as the late examples of Macedonian tombs date from Hellenistic times (3rd and 2nd centuries BC), indicating the continuity of customs and of life in the city even when the kingdom was on the wane. Tombs from the period of Roman rule, after 150 BC when the Macedonian state was finally abolished, appear sporadically over a wide area among the earlier monuments, as well as further north, as far as the river Aliakmon.

The monuments are numerous and only some of them can be visited today (royal tombs, "Rhomaios tomb"). As a result the visitor has only a limited picture and rudimentary knowledge of this important site. The many finds from the cemetery are kept in the archaeological storerooms of the Vergina excavation and in the Veroia Museum except for some from the royal tombs which are exhibited under the Protective Shelter over the royal tombs at Vergina.

The tumuli cemetery

The area between the modern villages of Vergina and Palatitsia is covered by the vast tumuli cemetery, in which over three hundred prehistoric burial mounds dating to the Early Iron Age (1000-700 BC) have been found, while a smaller number is dated to Hellenistic times (4th-2nd century BC). The site and form of the early settlement (or settlements) to which the large cemetery belongs are not yet clear. Nevertheless the copious and varied grave goods indicate a strictly stratified society and a high cultural level. This early settlement must have been succeeded by the ancient city which is now being excavated. It is the same city in which important monuments, such as the palace, the theatre and the sanctuaries, were built from Classical to Hellenistic times.

The tumuli cemetery spread over the centuries, mainly westwards to where the present village of Vergina now stands. It was obviously used by the inhabitants of the ancient city from Archaic into Hellenistic times, as the plethora of precious grave goods indicates. One of the last early tombs from the Iron Age, to the west, was found under the fill of the Great Tumulus, in its east part (7th century BC).

The Early Iron Age section of the cemetery has distinctive traits: each tumulus includes more than one burial (from five to fifteen), male and female, suggesting that it served a family. The tumulus itself, a small mound of red earth heaped over the

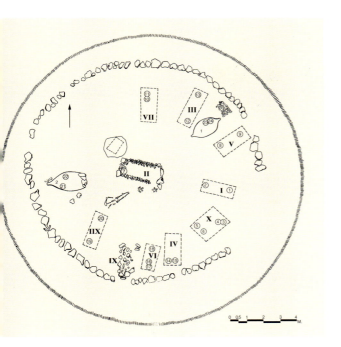

Fig. 40. The prehistoric tumulus was usually formed from red earth and defined by a circle of large stones. It evidently covered several tombs in most cases (Iron Age)

graves, is encircled by a row of stones. Very little skeletal material has survived, on account of soil conditions, but the sex of the dead can usually be deduced from the grave goods — clay vases, weapons and jewellery.

Presumably the tumulus covered graves of one family that included members with military and religious authority. Next to these are clusters of tumuli which might represent larger social groups, perhaps lin-

Figs 41-42. A rich female burial from the tumuli cemetery. Two vases had been placed near the head and a third at the feet. Elaborate bronze jewellery (spectacle fibulae, bracelets, a belt with bosses) adorned the deceased (8th century BC).

eages. The most important objects accompanying the dead are weapons. Virtually all are of iron and outstanding among them are the swords and knives. The female jewellery (fibulae, hair rings) is of bronze; interesting are the bronze spectacle fibulae as well as the peculiar hair ornaments. In some cases the jewellery is not merely for adornment but seems to have a religious or symbolic content (diadem, triple axe). Noteworthy among these pieces is the "crown", a broad band diadem decorated with circles inscribed with a cross, a solar symbol, which motif permits us to interpret the object as a "crown" (Vergina). Ubiquitous are the clay vases, such as oenochoai, Protogeometric skyphoi and bowls, pointers to local production and trade as well as to the more secure dating of the graves.

The manner of inhumation and the kinds of grave goods reveal the role of the family, the military character of the society, the religious ideology, albeit hinted at, as well as worship with symbols such as that of the sun.

Research has shown that the burials from the periods after the 7th century BC were dispersed, mainly to the west. These include rich graves from the sixth and fifth centuries BC, such as those in the vicinity of the so-called "Rhomaios tomb".

Fig. 44. Bronze object in the form of a triple double-axe, from a female burial in the tumuli cemetery.

The Classical stele of the 5th century BC, with the representation of a young hoplite, found in the Great Tumulus, also dates from this period. The clusters of opulent Macedonian tombs, as well as, primarily, the simple cist graves or brick-built graves, belong to the later (Classical and Hellenistic) section of the cemetery, the limits of which are still unknown. Recent research has shown that these outstanding monuments represent a major chapter in Macedonia's history.

Archaic, Classical and Hellenistic graves

The most important Archaic and Early Classical graves were found and excavated in the environs of the "Rhomaios tomb". Spanning over a century (second half of 6th-5th century BC), their grave goods revealed a prosperous world and a hierarchical society that traded with southern Greece, Attica and Ionia.

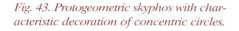

Fig. 43. Protogeometric skyphos with characteristic decoration of concentric circles.

Clay vases from Attic, Corinthian and Ionian (East Greek) workshops, as well as local wares indicate the mercantile activity at this time. However the most impressive finds are the jewellery and the cerements of the dead (gold earrings and necklaces, silver sandals, embroidered shrouds), that attest great wealth. The bronze vessels, the toys and the clay figures of the gods accompanying the dead reveal a life full of significant events and intense religious devotion. Outstanding for their religious content and artistic expression are the numerous terracotta figures of korai and male daemons, from the early fifth century BC, that were found in one of these graves. Although the graves of the fifth century BC were found plundered, the clay vases, most of them Attic, show the affluence of the Macedonians in this period and their relations with southern Greece.

Rescue and systematic excavations brought to light numerous graves from Archaic and Classical as well as from Hellenistic times. The area around and below the Great Tumulus was used intensively during these years. Hellenistic graves were located on the east outskirts of the village

Fig. 46. Terracotta female head from a Late Archaic grave in the cemetery.

of Vergina, as well as inside the area of the prehistoric tumuli cemetery, indicating the continuous use of the burial ground for nearly one thousand years. Weapons, jewellery and, above all, clay vases set their seal on the burial procedure and attest the continuity of society in the region.

Fig. 45. Pair of gold earrings from an Archaic grave in the cemetery.

Fig. 47. The large cist grave at Palatitsia. Detail of a bird, from the painted zone that runs round the walls.

The cist grave at Palatitsia

In 1982 a large cist grave covered by a relatively large tumulus was discovered at the eastern edge of the ancient cemetery lying between Vergina and Palatitsia. Because the grave, built of poros stone, is much bigger than usual it is covered by a double row of large poros blocks. Inside the grave is a spacious twin chamber with two large pillars with separate capitals along the main axis. A stone-built base-bench in the north section received the bier. The walls of the tomb are adorned with a painted band of running vegetal spiral, with a unique small, painted bird enriching the decoration on the west side. Although the tomb was found looted, the very few remaining finds date it *circa* 350 BC and reveal that the occupant was a woman, who was probably interred with rich grave goods.

Fig. 48. The large cist grave at Palatitsia with the double row of capstones (c. 350 BC).

THE GREAT TUMULUS - THE ROYAL TOMBS

The man-made hillock that covered some of the royal tombs at Aegae is known as the Great Tumulus. It is 13 m high and approximately 100 m in diameter. This large tumulus had been created in antiquity after the retreat of Pyrrhus and his mercenaries, following the looting of the tombs of the ancient city by Galatian invaders (274/3 BC). It was raised over the tombs and the small tumuli sixty years after their creation.

The excavation of the Great Tumu-

...In this year Alexander, succeeding to the throne, first inflicted due punishment on his father's murderers, and then devoted himself to the funeral of his father. He established his authority far more firmly than any did in fact suppose possible...

Diod. Sic. XVII, 2

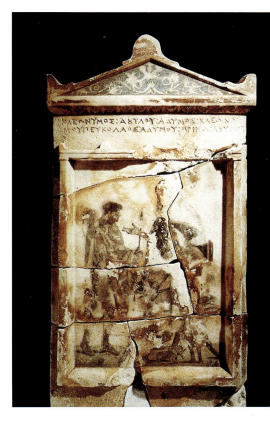

Fig. 49. Impressive painted stele of the 4th century BC, from the fill of the Great Tumulus. Its inscription preserves the names of four generations of the same family: Kelonymos son of Akylos, Adymos son of Kleonymos, Peukolaos son of Adymos, Krino daughter of Adymos.

lus (1976-1980) brought to light many important funerary monuments, which were considered a landmark for the history and archaeology of Macedonia. In 1992 after its completion, a large protective shelter with the same external form as the tumulus was constructed over these important monuments. So today's visitor has a verisimilar image of the great funerary mound that marked the position of the tombs. In 1976 many grave stelai of the fourth and the early third century BC were unearthed. Most of them are painted and preserve the name of the deceas-

Fig. 51. Plan of the shelter and the monuments it protects.

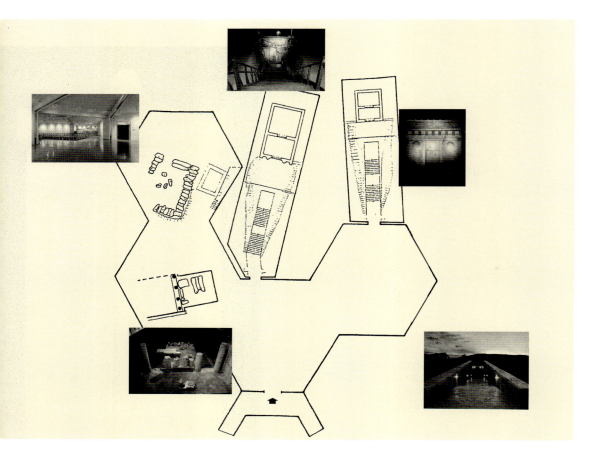

ed, very significant evidence for our knowledge of ancient Macedonian society. The discovery in 1977 of a large unplundered Macedonian tomb (**The tomb of Philip**) was momentous indeed, since it proved to be the last resting place of King Philip II of Macedon, the distinguished military and political figure of the fourth century BC. The king was assassinated in the theatre at Aegae during the celebration of his daughter Cleopatra's marriage, thus opening the way for Alexander's reign. He was buried hurriedly in the large tomb, along with his rich grave goods.

Fig. 52. The Great Tumulus at Vergina. The huge volume of earth was held in place by a complex network of rubble- masonry walls two-thirds up its height. The sepulchral monuments were in the south part of the periphery of the mound.

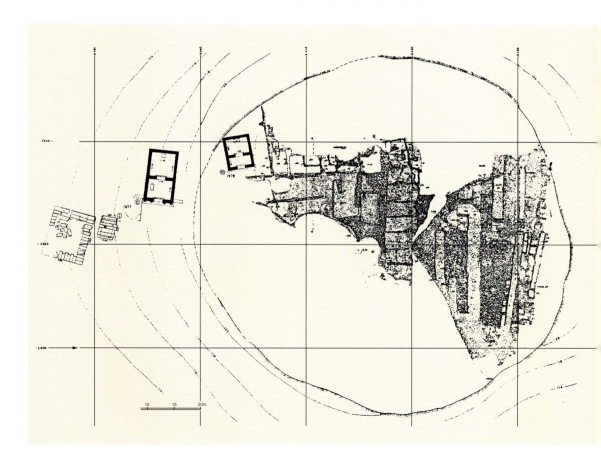

Figs. 53-54. Ivory heads from the decoration of the chryselephantine bier in the chamber of Philip II's tomb.

Fig. 55. Maquette of the ensemble of royal tombs (in the background) and the heroon (in front).

Precious weapons and costly silver and bronze vessels accompanied the deceased. The chryselephantine bier was embellished with an ivory relief frieze, an exceptional work of miniature carving. In the antechamber was another burial, of a young woman, possibly Philip's last wife, Cleopatra. Her bones were wrapped in cloth of gold and deposited in a gold larnax. Another chryselephantine bier completed the rich furnishings of the tomb, while the precious gold wreaths and diadems bear witness to the inordinate wealth of the dead. However, the main feature of the tomb is the large mural on its Doric facade. Depicted on a frieze 5.60 m long is a many-figured hunting scene, set within a verdant grove, remarkable for its composition and coloration.

Fig. 56. Ivory figure of a Muse playing a lyre. From the decoration of the chryselephantine bier in the chamber of Philip II's tomb.

Fig. 57. The large gold larnax found inside the marble sarcophagus in the chamber of Philip II's tomb. The dead king's bones were placed in the precious casket after his cremation. The lid is decorated with the large sixteen-rayed star, while on the sides are bands of lily, palmette and rosette ornaments.

Fig. 58. The small gold larnax from the chamber of Philip II's tomb, in which the bones of the young woman buried together with the king were kept.

Fig. 59. The gold-embroidered purple cloth in which the bones of the dead female were wrapped and placed in the small gold larnax. A large vegetal motif with birds recalls the large vegetal compositions on the diadem found inside the larnax, as well as other contemporary pieces of jewellery.

Fig. 60. Reconstruction drawing of the facade of Philip II's tomb. The wall-painting of the hunt, adorning the upper part, is the work of a major painter of the second half of the 4th century BC. (Drawing by G. Miltsakakis).

Fig. 61. Detail from the central part of the wall-painting of the hunt, on the facade of Philip II's tomb. The young huntsman on horseback perhaps depicts Alexander himself.

Fig. 62. Detail from the left part of the wall-painting of the hunt, on the facade of Philip II's tomb. The boldly drawn hunter turned diagonally, with his back to the viewer, denotes the achievements of the major painters in Classical Antiquity.

Fig. 63. View of the exhibition. In the interior of the shelter of the royal tombs.

Fig. 64. The gorytos or quiver, revetted with an alloy of gold and silver, decorated with repoussé battle scenes. From the antechamber of Philip II's tomb.

Fig. 65. The iron and gold-embellished cuirass from the chamber of Philip II's tomb.

Fig. 66. Silver strainer for decanting wine. From the chamber of Philip II's tomb.

Fig. 67. Bronze lantern from the chamber of Philip II's tomb. Below the point where the double-hoop handle joins the vessel is a wreathed head of Pan.

Fig. 68. Silver kalyx from Philip II's tomb.

Figs 69-70. Silver oenochoe from Philip II's tomb. At the base of the handle a remarkably expressive face in relief.

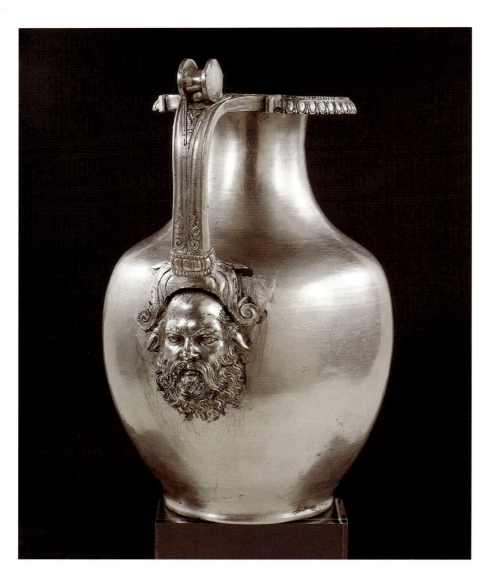

Fig. 71. View of the interior of the tomb of "Persephone". Left, the abduction of the young goddess by Pluto. Right, a sorrowful figure, perhaps Persephone's mother, the goddess Demeter. The work of a distinguished painter of the mid-4th century BC, most probably Nikomachos.

In the same year (1977) another funerary monument was brought to light, the so-called "**Tomb of Persephone**". The identity of its female occupant is unknown. Although it had been rifled, wonderful wall-paintings with subjects inspired by mythology are preserved in its interior. Outstanding among these is the scene of the Abduction of Persephone by Pluto, god of the Underworld, while on the narrow side is a representation of the goddess De-meter seated on a rock, grieving the loss of her daughter. Another three female figures are depicted on the other long side. The wall-paintings with their accomplished drawing and lovely colours are works of considerable artistic merit, that justifies their attribution to a painter of stature, perhaps Nikomachos.

In 1978 another Macedonian tomb was discovered next to that of Philip II. A little smaller and somewhat later, it too was unplundered. The

Fig. 72. Detail from the wall-painting in the "Tomb of Persephone". Pluto, king of the Under World.

Fig. 73. Detail from the wall-painting in the "Tomb of Persephone". Persephone.

Fig. 74. Detail from the wall-painting in the "Tomb of Persephone". Persephone's terrified female companion watches the abduction.

Fig. 75. Detail from the wall-painting in the "Tomb of Persephone". Demeter.

Fig. 76. The "Prince's tomb" is adjacent to Philip II's tomb. Of smaller dimensions and later in date, there are relief and painted shields on the facade, rather than engaged columns. The upper surface bore a panel painting, executed on perishable material (hide or wood), which has left only scant traces.

grave goods included precious vessels and weapons, and an ivory-and-gold bier. The dead was a youth about 14 years old, who must have been a member of the royal family, for which reason the tomb was conventionally named the "**Prince's tomb**". His bones were kept in a silver urn (hydria), crowned by a gold wreath and placed upon a stone podium. A narrow frieze with the subject of a chariot race, painted in

Fig. 77. The "table" on the west wall of the chamber of the "Prince's Tomb", with the silver cinerary urn. A gold wreath over the neck of the vessel rest on its shoulder.

Fig. 78. Detail from the wall-painting with the chariot race, which runs all round the walls of the antechamber of the "Prince's tomb". The excellent preservation of the original colours is impressive indeed, as is the way in which variations of the chariots and their drivers are repeated.

lovely colours, decorates the interior walls of the antechamber.

The splendid assemblage of monuments in the Great Tumulus is completed by another two, badly destroyed but nonetheless highly significant.

Next to the "Tomb of Persephone" are the ruins of a poros building above ground, a "**heroon**" in honour of the dead in the adjacent Macedonian tomb, that is Philip's.

Fig. 79. Part of the interior of the chamber of the "Prince's tomb", after it was opened. The precious silver vessels can be seen on the floor.

Fig. 80. The silver cinerary urn from the "Prince's tomb". The bones of the young man were placed in this large hydria, which was crowned with a gold wreath.

A few metres away from these large sepulchral monuments, on the periphery of the Great Tumulus, the remains of a third Macedonian tomb, which had been badly destroyed, were uncovered in 1980. This tomb, known as the "**Tomb of the Free-standing Columns**", must have been constructed around 300 BC and was richly furnished as apparent from the few but impressive finds (ivory heads etc.). The facade was arranged with four free-standing Doric columns.

The tombs in the Great Tumulus not only comprise an ensemble of important historical monuments but also a marvellous sample of the great painting of Classical antiquity. The wall-paintings and the funerary stelai bear witness to the techniques and splendid achievements of art at that time. The monuments in the Great Tumulus, now under the protective shelter, are a veritable museum of ancient painting. Towards the end of 1997, the Hellenic Ministry of Culture inaugurated a selective exhibition of objects and grave goods from the tombs in the area, both large and small, under the shelter. Displayed are the gold lar-

nakes from the tomb of Philip, some weapons and vessels, maquettes and photographs of the monuments. Outstanding among the exhibits are the chryselephantine biers from Philip's tomb, unique works for their superb art and precious materials. It is appropriate to mention here the skills of the conservators who succeeded in restoring the form of these priceless pieces of furniture.

Fig. 82. The Great Tumulus. The "Tomb of the Free-standing Columns" (c. 300 BC).

The tomb with the throne, the so-called "Tomb of Eurydike"

A few metres east of the "Rhomaios Tomb" an unexpectedly large sepulchral monument with unusual features was brought to light in 1987. It proved to be particularly important, since the finds show that it is the earliest known Macedonian tomb. Although it was found despoiled the vase sherds collected, indicate that it was built around 340 BC and that the dead was a woman of high rank, perhaps a queen (Eurydike?). The peculiar manner of the tomb's construction belies the still preliminary stage in the creation of the type: the walls are double, the vault is boxed into a rectangular cist-shaped structure, the facade is completely undecorated, and has been covered entirely, including the entrance, by a wall. Moreover, the monument is very long (10.60 m) and wide (7.90 m), while the interior is arranged in two sections, the antechamber (2.50 × 4.48 m) and the chamber (5.51 × 4.48 m), 5.80 m high. The plundering of the monument was extensive and destructive. Nevertheless, from the remaining evidence in the chamber, impressive and important conclusions were drawn. The back blind wall of the chamber is arranged as an Ionic facade of a Macedonian tomb, in which the colours, the stucco and the architectural details are preserved completely. In the middle of the wall there is a false door, framed by two pairs of engaged Ionic columns, between which are false windows made of stucco. The entablature with the three-taeniae epistyle, the frieze of flowers and the cornice, all in vivid colours, are truly impressive.

In front of this wall stands an un-

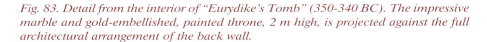

Fig. 83. Detail from the interior of "Eurydike's Tomb" (350-340 BC). The impressive marble and gold-embellished, painted throne, 2 m high, is projected against the full architectural arrangement of the back wall.

Fig. 84. The painted panel on the back of the marble throne in the chamber of "Eurydike's Tomb" depicts the gods of the Underworld, Pluto and Persephone, in divine majesty upon a quadriga.

precedented find: a richly embellished marble throne 2 m high. Decorated with reliefs, gilded flowers and animals on all visible parts, as well as with statuettes of korai, it has a superb painted "panel" on the back. Depicted within a lavishly decorated frame is a couple in frontal pose on a quadriga drawn by two white and two brown horses. The man holds a goad, the woman at his side holds a sceptre. The rendering of the figures permits their identification as the monarchs of the Underworld, Pluto and Persephone, an important divine *hypostasis* for worship in Macedonia, where the belief in life after death created special manifestations in religious life and inspired important works of art.

Fig. 85. Detail from the impressive architectural arrangement of the back wall of the chamber of "Eurydike's Tomb". All the elements of the Ionic style are rendered with exquisite accuracy, in relief and painted on the white marble stucco.

Fig. 86. The facade of the "Rhomaios tomb", with the engaged Ionic columns (c. 300 BC).

The so-called "Rhomaios Tomb"

The elegant Macedonian tomb seen by the visitor as he walks up towards the palace is one of the earliest known monuments on the archaeological site. It was excavated and studied by Professor K. Rhomaios, for which reason it is conventionally known as the "Rhomaios Tomb".

A two-chambered subterranean building of pleasing proportions, its impressive facade is dominated by the central marble portal (h. 3.14 m), framed by four elegant, engaged, Ionic columns. On the entablature above the epistyle, a narrow taenia-frieze was painted with a row of splendid flowers. The walls of the antechamber and the chamber, at the springing of the vault (height 2.22 m), were decorated with similar palmette bands. Unfortunately very few elements of this decoration are preserved and it is difficult for today's visitor to appreciate their form.

The dimensions of the two chambers are harmonious. The antechamber is 4.56 m wide and 2.50 m. long, while the chamber is of the same width and 4.56 m long. The marble throne is the most important feature in the tomb's interior. This impressive piece of furniture is decorated with small painted friezes in its lower part and with sculpted sphinxes on the armrests. A small marble footstool completes the picture and the function of the tomb's interior. In addition to the throne there is also a stone bench, which perhaps served as a base for the bier or the catafalque. The tomb was looted some time in the past, but it can be dated on the basis of its architectural features to around 300 BC.

Fig. 87. Reconstruction drawing of the facade of the "Rhomaios tomb". (Drawing by Ch. Lefakis).

Fig. 88. The interior of the "Rhomaios tomb". The stone throne and the bench.

The tomb next to the Vergina Cultural Centre (today's Town Hall)

This large, badly destroyed tomb stands on the east side of the village of Vergina, next to the Cultural Centre. Although ruined, several characteristic traits have survived so that a graphic reconstruction is possible. Part of the facade is preserved with the four engaged Ionic columns framing the opening to the tomb, an opening without the usual marble portal. On the contrary, there was a large marble portal between the antechamber and the chamber, as the few fragments found inside attest.

The tomb comprises two parts, a chamber and a particularly larger antechamber, an arrangement that is rather unusual in Macedonian tombs. A very narrow bench runs round the walls of the antechamber, while the floor of both the antechamber and the chamber is of thick red stucco. The dead was male and held military office, as inferred by part of an iron helmet found inside the tomb. Although the tomb was found destroyed and looted, the scant finds date it just after 300 BC. It thus represents the beginning of the Hellenistic period for this kind of funerary monument.

Fig. 89. The Macedonian tomb adjacent to the Cultural Centre at Vergina (today's Town Hall). The arrangement of the two large sections is visible, as well as part of the facade with the engaged Ionic columns.

The Macedonian tombs on the Bella farm

In 1981-1982, a short distance southwest of the Macedonian tomb discovered by L. Heuzey in the mid-nineteenth century, three looted Macedonian tombs and one un-looted cist grave, the sole grave good in which was a gold wreath,

Fig. 90. Tombs on the Bella farm. The Macedonian one-chamber tomb with the wall-painting on the facade. The marble throne is visible in the interior (first half of 3rd century BC).

were discovered under a very low mound. These tombs were named after the owner of the farm on which the tumulus was located (Bella farm). The mound that covered them was barely visible on account of modern ploughing and so we have only a vague picture of its original aspect. All the tombs in it date from the third century BC and constitute a representative sample of Macedonian tombs from this century, of the architecture in general and the monumental painting from the first and the middle Hellenistic phase.

The Macedonian tomb with the throne is the earliest monument in the Bella group. A single space, with no particular architectural articulation of the facade (h. 4.80 m), it imposes by virtue of the harmonious dimensions of its interior (3.50 × 3 m), the large marble portal and the painted decoration on the facade. The simple composition covers the surface of the wall above the lintel. At the centre of the representation is the figure of a warrior in cuirass. He leans on his spear and gazes into the distance, while his body, depicted in splendid colours, turns en face towards the viewer. The central figure, in military attire who can presumably be identified with the dead in the tomb, is flanked left by a slender female who offers him a wreath, and right by a male, also a warrior, sitting upon shields and observing the other two. Perhaps he is Ares the god of war, or perhaps another symbolic figure is depicted. The artist's desire to express and enhance martial valour and the glory of the dead is obvious. At the same time, despite the simple composition, he moulds these three figures through clever use of colours.

The tomb's plain interior is dominated by the marble throne at the centre of the wall opposite the door.

Fig. 91. The marble throne from the one-chamber Macedonian tomb. The back is painted on the wall.

Its back is painted in red on the wall's surface, while the piece of furniture itself was decorated in various colours and glass occuli on the volutes of the legs. The stone casket that was found in the tomb and which contained the probably precious ossuary of the dead, must have been placed upon this throne. The second Macedonian tomb in the Bella group should be dated in the second half of the third century BC. Excavation data reveal that it

Fig. 92. Detail from the wall-painting on the facade of the one-chamber Macedonian tomb on the Bella farm. The young man, perhaps a personification of Ares or Polemos (War), is depicted on the viewer's left, seated on a pile of shields.

Fig. 93. Detail from the wall-painting on the facade of the one-chamber Macedonian tomb on the Bella farm. The dead, at the centre of the representation, in full martial panoply, leans triumphantly on his spear.

Fig. 94. Detail from the wall-painting on the facade of the one-chamber Macedonian tomb found on the Bella farm. The slender female figure in the right part of the representation, perhaps a personification of the deceased's Arete (Valour), offers the warrior a wreath.

was used twice, with corresponding building interventions. Two-chambered with a built dromos (l. 7.80 m), it has a complex architectural form on the facade: the entrance is framed by four engaged Doric columns in pairs, while a concave cornice on the characteristic pediment considerably enlarges the height of the facade. All the architectural members are coated in white marble stucco. The antechamber is narrow with stucco masonry on the walls below, while the projecting kosmophoros is painted to imitate polychrome marble. The ceiling is horizontal with plastered stone beams. A stone door, of which both leaves are preserved, connects the antechamber with the marble-tiled chamber proper. This is dominated by the large stone sarcophagus in the form of a bier, with rich polychrome painted and relief ornaments. The covering of the sarcophagus imitates the palliasses of the bier, which are rendered in red, and on both narrow sides are stone pillows covered with figures such as Nikai. Outstanding too are the miniature representations of Dionysiac themes on the panel of the bier. Depicted is the reclining Dionysos whom a leopardess softly approaches. On the walls of the main chamber the taenia of the kosmophoros with painted imitation marble again separates the surfaces in white and ochre.

The third Macedonian tomb, which

Fig. 95. The stone bier-sarcophagus from the chamber of the two-chamber Macedonian tomb on the Bella farm. The bier bears relief decoration (second half of 3rd century BC).

faces west, must be the latest monument in the group on the Bella farm. Single-chambered, with no architectural articulation of the facade except the small decorative pediment above the door, it is dated at the end of the series of these sepulchral monuments in the late third century BC. Its interior contains a disproportionately large stone sarcophagus that occupies almost the entire area of the chamber (2.50 × 2.30 m). Fragments of many figurines show the kind and character of the burial it housed, which could be ascribed to some female.

The Macedonian tomb on the Bloukas field

In 1969 an undecorated, single-chamber Macedonian tomb was brought to light on the northern outskirts of the village of Vergina. Built of poros stone, like all the other Macedonian tombs, there is little articulation of the facade, while the inside walls were coated only with white stucco. The grave goods, mainly clay vases, showed that the tomb must have been built around 200 BC and was used for a further fifty years. It is considered one of the latest examples of the category of Macedonian tombs, postdating even the group of tombs on the Bella farm.

TIME CHART

359 BC	Philip crowned king
357 BC	Philip's marriage to Phila Siege and conquest of Amphipolis Philip's marriage to Olympias The Conquest of Pydna
356 BC	The Birth of Alexander
354 BC	The Conquest of Methone
348 BC	The Conquest of Olynthos
338 BC	The Battle of Chaeronia Peace with the Greek city-states
337 BC	The Corinth Confederacy
336 BC	The Assassination of Philip at Aegae Alexander crowned king
335 BC	Alexander in the Balkans
334 BC	The Campaign against Persia
334 BC	The Battle of Granicus The Fall of Miletos
333 BC	The Battle of Issus
332 BC	The Siege of Tyre
331 BC	The Founding of Alexandria in Egypt
331 BC	The Battle of Gaugamela Alexander in Susa
330 BC	In Persepolis
329 BC	In River Iaxartes
328 BC	In Bactria
327 BC	In River Indus
323 BC	In Babylon
323 BC	Death of Alexander
319 BC	Death of Antipater
317 BC	Death of Eurydike-Adaia and Philip Arrhidaios
316 BC	Death of Olympias
304 BC	Demetrios in Athens
301 BC	The Battle of Ipsos
300/299 BC	Alliance between Ptolemy and Lysimachos. Alliance between Demetrios and Seleucos
297 BC	Death of Cassander
294 BC	Demetrios Poliorcetes in Athens. Demetrios declared king
287 BC	Demetrios expelled from Macedonia
287 BC	Macedonians declare Pyrrhus king
285 BC	Lysimachos conquers Macedonia
282 BC	The Battle of Kouropedion Lysimachos slain
277 BC	The Battle of Lysimachia
276 BC	Antigonos Gonatas king
275 BC	Pyrrhus invades Macedonia Galatians plunder the tombs at Aegae
272 BC	Antigonos returns to Macedonia
257 BC	Demetrios co-regnant with Antigonos
240/39 BC	Death of Antigonos. Succeeded by Demetrios II
229 BC	Death of Demetrios II
227 BC	Antigonos Doson king
222/1 BC	Death of Antigonos Doson Philip V king
179 BC	Death of Philip V Perseus king
168 BC	The Battle of Pydna
167 BC	The Dissolution of the kingdom

HOUSE OF THE TEMENIDS

Karanos	Gauanes
Koinos	Aeropos
Tyrimmas	Perdiccas

HOUSE OF THE ARGEADS

Perdiccas I		Pausanias	393 BC
(Argaios)		Amyntas II	393 BC
Philip I		Amyntas III	392-370 BC
Alketas		Argaios	390 BC
Amyntas I	497 BC	Alexander II	370-368 BC
Alexander I Philellen	497-454 BC	Ptolemy Alorites	368-365 BC
Perdiccas II	454-413 BC	Perdiccas III	365-359 BC
Archelaos	413-399 BC	Philip II	359-336 BC
Orestes	399-396 BC	Alexander III	336-323 BC
Aeropos II	396-393 BC	Alexander IV	323-310 BC

HOUSE OF THE ANTIPATRIDS

Cassander	315-297 BC	Antipater	297-294 BC
Philip IV	297 BC	Alexander V	297-294 BC

HOUSE OF THE ANTIGONIDS

Demetrios Poliorcetes	294-288 BC	Antipatros	
(Pyrrhos)	288/7 BC	Antigonos II Gonatas	277-239 BC
Lysimachos	287-281 BC	Demetrios II	239-229 BC
Ptolemy Keraunos	281-279 BC	Antigonos III Doson	229-222 BC
(Anarchy)		Philip V	222-179 BC
Meleagros		Perseus	179-168 BC

SELECTED BIBLIOGRAPHY

AErgoMak (=*Το Αρχαιολογικό Έργο στη Μακεδονία και Θράκη*), 1987-1997. Articles about the monuments at Vergina by the excavators S. Drougou, Ch. Saatsoglou-Paliadeli, P. Faklaris, A. Kottaridou, E. Tsigarida.

M. Andronikos et al., *Το ανάκτορο της Βεργίνας*, 1961.

M. Andronikos, *Βεργίνα I. Το νεκροταφείον των τύμβων*, 1969.

M. Andronikos, Ανασκαφή στη Μεγάλη Τούμπα της Βεργίνας, *AAA* IX (1979), 123 ff.

M. Andronikos, *Βεργίνα. Οι βασιλικοί τάφοι και οι άλλες αρχαιότητες*, 1984.

M. Andronikos, *Βεργίνα II. Ο τάφος της Περσεφόνης*, 1994.

S. Drougou, Το ύφασμα της Βεργίνας, *Αμητός, Τιμητικός τόμος για τον καθηγητή Μανόλη Ανδρόνικο*, 1987, 303 ff.

S. Drougou - Ch. Saatsoglou-Paliadeli - P. Faklaris - A. Kottaridou - E. Tsigarida, *Η Μεγάλη Τούμπα* (archaeological guide), 1994.

P. Faklaris, Πήλινες μήτρες, σφραγίδες και ανάγλυφα αγγεία από τη Βεργίνα, *ADelt* 38 (1983), Meletes, 37 ff.

N.G. Hammond - G.T. Griffith, *A History of Macedonia*, I-III, 1972-1979.

L. Heuzey - H. Daumet, *Mission archéologique de Macédoine*, 1876.

Μακεδονία. Από τον Φίλιππο Β' έως τη ρωμαϊκή κατάκτηση (M. Andronikos) (ed. R. Ginouvès) 1993 (bibliography).

D. Pandermalis, Ο νέος μακεδονικός τάφος της Βεργίνας, *Μακεδονικά* 12 (1972), 147 ff.

D. Pandermalis, Η κεράμωση του ανακτόρου της Βεργίνας, *Αμητός, Τιμητικός τόμος για τον καθηγητή Μανόλη Ανδρόνικο*, 1987, 579 ff.

D. Pandermalis, Beobachtungen zur Fassadenarchitektur und Aussichtsveranda im hellenistischen Makedonien in *Colloquium "Hellenismus in Mittelitalien" (Göttingen 1975), AbhGöt* 91.1 (1976), 387-395.

K. Rhomaios, *Ο μακεδονικός τάφος της Βεργίνας*, 1951.

Ch. Saatsoglou-Paliadeli, *Τα επιτάφια μνημεία από τη Μεγάλη Τούμπα της Βεργίνας*, 1984.

Ch. Saatsoglou-Paliadeli, Ευρυδίκα Σίρρα Ευκλεία, *Αμητός, Τιμητικός τόμος για τον καθηγητή Μανόλη Ανδρόνικο*, 1987, 733 ff.

E. Tsigarida, Χρυσό στεφάνι μυρτιάς από τη Βεργίνα, *Αμητός, Τιμητικός τόμος για τον καθηγητή Μανόλη Ανδρόνικο*, 1987, 902 ff.

Y. Velenis, Τεχνικές στο ανάκτορο της Βεργίνας, *Μνήμη Μανόλη Ανδρόνικου*, 1997, 25 ff.

USEFUL INFORMATION

Vergina is located 8 km southeast of Veroia. There is a local bus service between Veroia and Vergina. Veroia is linked to Thessaloniki by road and rail with regular bus and train services (information about buses KTEL Imathia and trains OSE). There is also a direct link between Veroia and Athens.

There are parking facilities at the archaeological site and southeast of the village. There is a bookstore and a refreshenent stand in the garden of the royal tombs. There are restaurants and cafes in the village of Vergina. Accommodation is available in Veroia and Vergina.

The archaeological site includes the royal tombs (ticket office) and the palace-theatre (ticket office). Since November 1997, there is an exhibition of finds from the tombs, under the protective shelter. The finds from the rest of the archaeological site are not exhibited.

Tel. archaeological site of Vergina   23310-92347
Tel. Community of Vergina   23310-92337
Tel. Vergina excavations (Aristotle University of Thessaloniki)   23310-92795

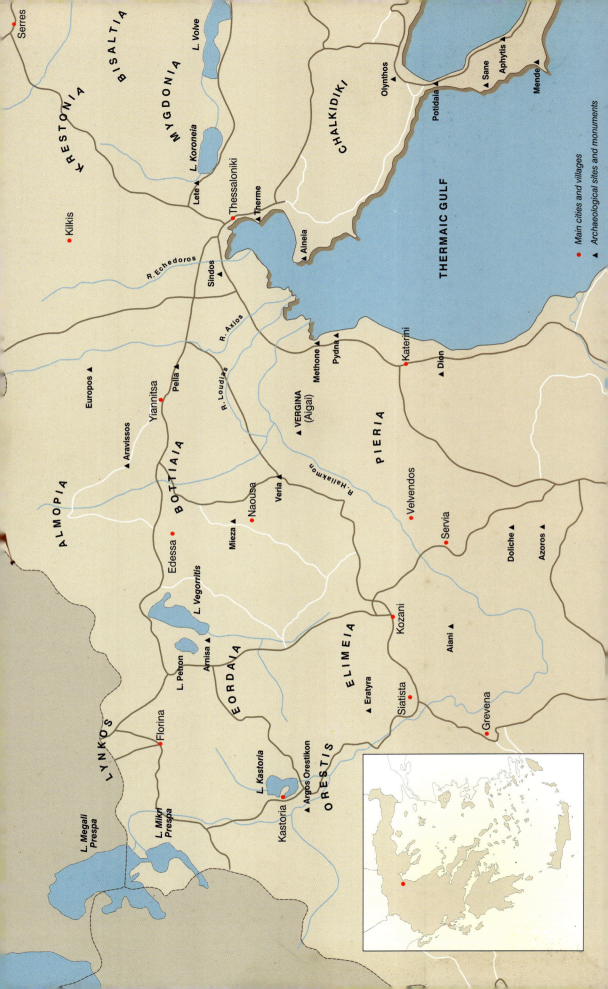

Serres

KRESTONIA

BISALTIA

L. Volve

MYGDONIA

L. Koroneia

CHALKIDIKI

Olynthos

Sane

Aphytis

Potidaia

Mende

Kilkis

Lete

Thessaloniki

Therme

Ainela

THERMAIC GULF

R. Echedoros

Sindos

Main cities and villages

Archaeological sites and monuments

R. Axios

Katerini

Europos

Methone

Pydna

Dion

Yiannitsa

Pella

R. Loudias

Aravissos

VERGINA
(Aigai)

PIERIA

ALMOPIA

BOTTIAIA

Naousa

Veria

R. Haliakmon

Velvendos

Mieza

Servia

Doliche

Azoros

Edessa

L. Vegorritis

Kozani

Aiani

Arnisa

L. Petron

ELIMEIA

EORDAIA

Eratyra

Siatista

Grevena

Florina

ORESTIS

LYNKOS

L. Kastoria

Argos Orestikon

L. Megali
Prespa

L. Mikri
Prespa

Kastoria

PROCEEDINGS

SPIE—The International Society for Optical Engineering

Diamond Optics

Albert Feldman, Sandor Holly
Chairs/Editors

16–17 August 1988
San Diego, California

Sponsored by
SPIE—The International Society for Optical Engineering

Cooperating Organizations
Applied Optics Laboratory/New Mexico State University
Center for Applied Optics Studies/Rose-Hulman Institute of Technology
Center for Applied Optics/University of Alabama in Huntsville
Center for Electro-Optics/University of Dayton
Center for Optical Data Processing at Carnegie Mellon University
Georgia Institute of Technology
Institute of Optics/University of Rochester
Optical Sciences Center/University of Arizona

Published by
SPIE—The International Society for Optical Engineering
P.O. Box 10, Bellingham, Washington 98227-0010 USA
Telephone 206/676-3290 (Pacific Time) • Telex 46-7053

Volume 969

SPIE (The Society of Photo-Optical Instrumentation Engineers) is a nonprofit society dedicated to advancing engineering and scientific applications of optical, electro-optical, and optoelectronic instrumentation, systems, and technology.

The papers appearing in this book comprise the proceedings of the meeting mentioned on the cover and title page. They reflect the authors' opinions and are published as presented and without change, in the interests of timely dissemination. Their inclusion in this publication does not necessarily constitute endorsement by the editors or by SPIE.